How to Strategically Position Projects (Or What PM Courses Don't Teach)

J.Barré

Published by Always Grow Publishing, 2024.

While every precaution has been taken in the preparation of this book, the publisher assumes no responsibility for errors or omissions, or for damages resulting from the use of the information contained herein.

HOW TO STRATEGICALLY POSITION PROJECTS (OR WHAT PM COURSES DON'T TEACH)

First edition. April 18, 2024.

ISBN: 979-8227130013

Written by J.Barré.

Table of Contents

I dedicate this book to my beautiful, highly intelligent, and compassionate daughter.

Her immense curiosity and interest for my work and the world as a whole motivates me every day to be a better mother and consultant.

Love, mama xXx

Disclaimer

The concepts and theories shared in this book are based on the author's education and experience. They are not meant to replace existing standards, internationally recognized or otherwise, nor proclaim that all projects or contexts in which projects are conducted are the same as what the author has described.

Definitions were sourced as per the sources in the References & Resources section.

While inspiration for the majority of the diagrams illustrated in the book was drawn from existing content and best practices, each was adapted or redesigned by the author, to prevent any copyright infringement.

Contributions

I would like to acknowledge an invaluable contribution to the final version of the book by Jocelyn Lortie, a Project Manager (PM) extraordinaire, whom I had the pleasure of working with through a general election in Canada and also have had the honour of considering a mentor.

Merci, Jocelyn.

You can check out Jocelyn's profile on LinkedIn[1].

---

1. https://www.linkedin.com/in/jocelyn-lortie-7a10893/

Helping Women & Children in Need

A portion of book sales will go to the Wealthy Heart Foundation's programs to support women who are facing adversity and are determined to make a positive impact on the world, including carving a path of generational wealth for their children.

For more information, visit the website:

Wealthy Heart Foundation[1].

---

1. https://www.wealthy-heart-foundation.com/

Prologue

They say project management is part art and part science, but most of the teachings on project management focus on the 'science' part of it. We're taught about the Project Life Cycle, how to make plans and to estimate costs, but we're not taught how to deal with challenging situations, especially the people type, or to prevent these situations in the first place.

A lot of 'soft' aspects are key to delivering successfully on project objectives. Some of these aspects, when not planned for or managed, can downright jeopardize completion. In fact, close to 70% of (IT) projects fail (Endnote 1), and the top reasons given for this are not related to methodology per se. A lack of clear goals (37%) was the most commonly identified factor in a recent KPMG survey, followed by inadequate stakeholder engagement (25%), ineffective risk management (23%), and poor communication (21%). Other factors for project failure previously identified by the Project Management Institute (PMI) and other sources include:

- Misalignment with Organizational Objectives
- Poor Planning
- Lack of Executive Support
- Incomplete / invalid requirements
- Unclear or unrealistic expectations
- Scope creep
- Lack of resources

- Poor project management

None of the factors listed above are unpreventable. By applying diligence upfront and some rigour throughout, I'm convinced that every project can be fruitful.

And how can I, an "accidental project manager", infer that? Because I've been involved in more than 40 projects and led/managed almost all of them. I've almost seen it all when it comes to projects, and I've gotten pretty great results. I was so good at getting results for my clients, in fact, that I was asked to assume this role over and over!

Was it because I followed the methodology to a T and was always exact at estimating budget and timelines? No. Was it because I was the best at creating complex plans in MS Projects? No. While I was very skilled at the technical part of the job, it was being able to manage the "people aspects" of the project that was key for me.

In this book, you'll find out about those soft skills and get some tips that I've applied in my Project Management practice to get great results.

Definitions for key terms that you may not be familiar with are provided at the end of the document. Just click the hyperlink on the specific term to get to that section.

Project Lifecycle

On a Personal Note...

Why do I see myself as an "accidental" PM?

I didn't plan to make a career as a project manager. In fact, upon graduating from my MBA 20 years ago, I wasn't exactly sure what I wanted to do – I had a sense of what I was looking for and mostly knew what I didn't want, but mostly knew that I wanted to make a difference. I wasn't looking to just make money. I also knew that I wanted to do some type of consulting work and be my own boss. At first, I thought I'd go into organizational / operational design or improvement, maybe strategic planning.

As I was canvassing for contracts, after returning from three weeks in India and starting to feel the pressure to earn an income again, I fell upon this posting from an alum who then worked as a director in one of British Columbia's large health authorities (I lived in Vancouver at the time).

The Health Authority was looking for a PM who had experience with information systems and training, and who could also do a needs assessment and write reports. That piqued my curiosity.

I'd learned to complete environmental scans, develop business cases and reports during my Masters' so was looking forward to doing that more in the real world. I'd trained people on information systems and done a bit of project management in my last position. I didn't have my PMP yet, but I'd taken a few courses in project management during the MBA's professional development segment and found the domain interesting.

I interviewed and was selected simply based on the above and a personality fit. This is how I undertook my first projects.

I enjoyed it and exceeded my client's expectation, so I continued. I held PM roles on a contractual basis in my early and late career, and as a permanent employee for three different organizations in between my freelancing experiences.

But I also took on mandates in a broader consulting capacity throughout my career.

The 'persona' that seemed to stick to me most though was the PM one. Why? Because I was good at it, I suppose. I enjoyed being able to conceive and plan an approach based on the objectives to be met, and then orchestrating everything to get the job done. I couldn't really think of another role that I'd enjoy that would provide endless opportunities to learn about different topics, to work with a variety of people, to earn a decent income and continue growing it, etc, and allow some flexibility with my hours and extended vacations between contracts.

So it was accidental in the sense that I kind of 'fell into it' organically by securing my first contract and also not by – like many do to start off in project management – studying to become a certified PM. In all honestly, I didn't even know about the certifications until after I started calling myself a PM!!

A lot of what I needed to be good as a PM came naturally, but I also learned from doing the work and from different education/training programs I took along the way. I finally got PMP certified in 2011 after practicing as a PM for over five (5) years. To be honest, the PMBOK may have only taught me half of the knowledge I've built over the years.

Don't get me wrong, I'm not discouraging anyone from getting a PMP, PRINCE 2, or Agile certification, or all three. One or more certifications are needed for many jobs and contracts. Besides, knowing and understanding proper methodologies certainly helps better managing projects, especially if you are able to optimally use one method versus another or when to combine them. I personally favour a Waterfall-Agile hybrid approach for projects that involve technology with a large group of stakeholders and multiple business processes. This is because the waterfall approach allows to properly scope the business processes, plan activities, and complete deliverables in a manner that I find essential from a business/client perspective – e.g., gap-fit, communications and engagement plan, risk management plan, reporting, etc. Agile does not give the same type of attention to this, nor document certain aspects of the project work in the way that the business client and other higher ups need. Agile works well for IT and other project team members who do the work. Integrating the two methods, if done well, satisfies both all of the business and technical needs of the client.

So, I encourage everyone to learn about one or more recognized approaches/methodologies and take from these what you need for the specific project you're leading. But I'd discourage anyone from taking either of these as the only source (or 'book of knowledge') you should have to be successful in project management.

Practical experience, common sense, and also mentors (e.g., seasoned PMs or leaders who have sponsored many projects) can also provide a wealth of information that will serve a new PM well.

You'll find that I greatly emphasize the role of the Project Sponsor – also known as (AKA) 'Executive Sponsor', 'Business Sponsor', or 'Sponsor', but abbreviated to PS for our purposes. This role is closely intertwined with the PM's and has a significant impact on project success. I found over the years that many of my clients (PS included) misunderstood what the PM role really is and what really being a (good) PM entails.

So, it became important to me to help both sides identify what each can do to work together more effectively.

I hope that my book is one of many you will read to complement your knowledge, but also that you will take courses and workshops that will provide case studies and opportunities to apply what you're learning.

The best way to learn ultimately is by doing the work. So, get in there and get your hands dirty after you read this!!

For more information about the author, visit:

https://www.jube-consulting.com/

Introduction

(Or Let's Get Project Failure Out of the Way)

Earlier, I listed the most-commonly identified **Factors leading to project failure** without providing more details or ways to avoid them. As the rest of this book focuses on other takeaways that will help you strategically position projects and yourself as a project leader, the below section is for you if you're interested in why projects fail and what to do to prevent that from occurring on your own projects.

1- Misalignment with Organizational Objectives & Lack of Clear Goals

- *Why does this happen?*

Some projects start because someone with influence thought it would be a 'good idea' or 'really innovative' to (NAME AN INITIATIVE). However good the idea might be on its own, it may not be a good idea for your organization to invest time and money into it – all considered. This includes other priorities the organization may have at that moment.

Some will give the argument that "innovation is disruptive" and that you can't progress unless you have disruption, but they neglect the fact that innovation/disruption can also be very destructive if poorly planned (and executed) in the big, strategic scheme of things. Because it will either lack the resources (capital, human) to be completed properly, or it will 'steal' resources from other projects that will suffer and under-deliver for the organization – thus failing to meet strategic objectives.

If the project is initiated as a response to an opportunity or problem that is viewed as a priority by someone in the organization, but it is not aligned with a pre-defined strategic objective, it should not be given priority over other projects in the pipeline that were agreed-upon because they would support strategic objectives.

Of course, things in the organization's internal or external environment may change in the course of a Strategic Plan's lifecycle and it is up to the Board/Executive to determine whether such changes warrant a) the review of strategic objectives, or b) an exception to a project that does not appear to be aligned with strategic objectives to be added and prioritized over others. Projects that are meant to address an unforeseen, emerging crisis are examples or projects that should be considered for an exception.

- ***What to do to prevent Misalignment / Lack of Clear Goals?***

Every organization, big or small, should put organizational policies and procedures in place for project intake and approval by appropriate decision-makers. The complexity of the process – including the extent of steps/stakeholders involved in reviewing the project request – can be scaled to the size and project management maturity of the organization.

This can be executed through a proper pre-planning phase. The initial steps should, to a minimum, be as follows for the person who had the idea and/or who believes in the project (This could be the person who will be the PS during the project, or temporarily until it is approved):

1. Seek support from one or more member(s) of the Executive team
2. Obtain the authorization to mobilize resources to develop a project proposal, business case or both (in multiple steps) to justify the project and estimate resources required to execute it
3. Bring the resources needed, including a senior consultant and/or project manager, to establish the need for the project (i.e., what we're really trying to solve,) explore and develop options to solve it, cost things out, and evaluate the impacts and risks for the organization if nothing is done vs if what you're proposing is implemented – including consideration for the capabilities and time of the team(s) that will support project execution and/or who will be at the receiving end once the project is over.
4. Review the project proposal/ business case, and

approve or reject.

The time spent on properly defining what you'd like to do is basically a 'feasibility' stage, and it's absolutely worthwhile if done well. Not only will it help uncover information essential to making an educated decision about the project at this stage, but it will also bolster executive buy-in if the project is a strategic fit and presents more benefits than pitfalls. As most of you may know by now: "Failing to plan is like Planning to Fail" (Based on a quote by Benjamin Franklin). And most people don't want to fail!

For a project manager joining in at a later stage (most often during Initiation), it's essential to validate the project has received the proper approvals and that steps were not skipped to get the project 'presumably' approved. If there are gaps in the background information that is coming to you as a PM, make sure you somehow gather it. This may include going back to the drawing board. This is your only opportunity to say something before accepting the conditions of the project and becoming **responsible** for its delivery even if you had not agreed to the general scope, schedule, and cost provided to you at the time you took over. As a PM, it is your duty to explain to your PS why it is important that you ensure the project's alignment with (organizational) strategic objectives and, whether they exist or not, that objectives be developed for the project itself (i.e., what the project will do and deliver).

This assumes that the PS did not follow proper project management procedures, however seasoned sponsors who believe in the methodology will know to do this.

They will also know and support their PM in not moving to a subsequent phase in the lifecycle until the proper diligence is completed. At a minimum, the Planning Phase should not be started until the Initiation Phase has progressed to a point where Planning will not have to be entirely redone because previous steps were missed.

See Appendix A for a detailed version of the project lifecycle.

2- Unclear or Unrealistic Expectations

- *Why does this happen?*

This is tightly linked to *Lack of Clear Goals* – because much of the unrealistic expectations people have around projects stem from objectives not having been defined properly at onset.

Another reason why expectations may be unclear or unrealistic is that project constraints were not discussed. Every project is subject to constraints – whether Budget, Quality, Scope, and Time (Note: Constraints are also considered to be part of project objectives. I distinguish between the two for clarity). In most cases, sponsors look to get it all – that is, they want the most and best quality product or service delivered at the lowest cost and in the fastest time possible. But that's not achievable. One, or more of those, has to give; thereby the term 'constraint'. Project execution thus become constrained by the objectives that are of priority to the Client/PS. For instance, if the sponsor wants great quality and a certain scope but is limited in budget, the completion time could be extended to allow for the scope and quality reviews without having to add more resources or require overtime.

I've come across this scenario (and other such variations) more than once. For example, I led a project that entailed deploying a public health information system across 83 communities in a Canadian province for many years. 'Subscribing' to the system was left as a choice, but my clients wished to have as many communities as possible on board. So, we agreed to spend the required time to approach each community and gain 'adoption' - with diplomacy. We had a large scope, a decent but not unlimited budget, and a lot of time. The quality of the interactions with the communities and of our deliverables was important to our clients, and this was achieved by balancing Budget, Scope, and Time.

You will likely have come across the Triple Constraints model above. Additional examples and explanations can be found in Appendix B.

- ***What to do to prevent Unclear/Unrealistic Expectations?***

Developing and discussing objectives and what they mean with sponsors and key team members allows for different expectations and potential misunderstandings to come to light. This in turns provides an opportunity for the PM and project team to clarify related elements and to come to an agreement on what should delivered by the end of the project.

While the theory on developing objectives state that they should be SMART, I've found that it isn't always realistic to have such specificity for project objectives given the time and number of changes large and lengthy projects may undergo during their lifecycle. What matters most is having objectives that cover the breadth and depth of the project scope and deliverables, and to be specific about the later. This includes defining what is in/out of scope and precisely describing expectations of deliverables.

Developing objectives is an iterative process. Once you discuss desired outcomes and scope with the project team, you may find that you have too many or not enough objectives, or that the wording must be adjusted to encompass all that you need to do, or to accurately reflect what will be completed by the team. So, you should take another stab at it until all are satisfied that an alignment exists between objectives, scope, deliverables, and desired outcomes.

In addition, key performance measures for expected project outputs should be documented – to a minimum in a project charter or master plan – and be given sign off by leads and sponsor(s). This is to guide the project team as they develop products/services for the client(s).

The same goes for project constraints. Documenting them in a project charter is essential as human memory also fails. Project team members, as well as sponsors, will forget what was agreed-upon over time if not initially documented and/or updated when changes are initiated and authorized. This could cause issues related to expectations that have become unfounded. New team members will also appreciate this information when joining the project.

3- Ineffective Planning & Risk Management, Scope Creep, and Poor Project Management

- ***Why does this happen?***

The reason I put these two together is simple: Lacks or gaps in planning and risk management reflect poor project management. Scope creep indicates ineffective scope management, thus poor scope management. Planning, as well as managing risks and scope, are foundational tasks that a PM must complete on and throughout every project.

Some people are assigned a project because they are good leaders who were able to deliver on many programs or day-to-day activities within their organization – but they've never learned to put a plan together (whether a project schedule, communications plan, or risk management plan). That doesn't fly well when delivering projects. The larger a project is, the more challenging it becomes to execute without plans.

Of course, poor project management goes beyond ineffective planning, or risk management or yet scope management. Deficiencies in other project knowledge areas (see Section 1 for more details) can also lead to projects going off track. While ineffectiveness may be due to a lack of knowledge or understanding of the methodology, in many cases it is linked to not applying it with enough consistency, discipline, or rigour.

There are many types of project managers, and sometimes the person chosen just isn't the right fit for that specific organizational culture, or they may not have the chops for a given mandate.

There are also people who are better suited to the PM role than others. Some have great personalities; others don't. It takes guts, a good read on people, and some finesse to manage projects – because you're really managing people, not just plans. So people skills are necessary.

SIDEBAR: An Executive I met long ago would tell his staff "Suck it up, princess!" when they complained about the changes and work brought on by projects. He told me once with a dead serious tone: "Wouldn't it be good if there were no people to deal with when doing projects?" This baffled me – he'd lost sight of the reason of being for projects to begin with: The people are who with and why we do projects in the first place!

- ### *What to do to prevent Ineffective Project Management?*

Ideally, a systematic and sound application of project management methodology, as well as maintenance of related artifacts, would be sustained throughout the project.

For those who don't have a PM certification or training, or who were never exposed to project management tools, there is a wide array of resources online so there is no excuse not to know what project management is about.

To a minimum, a project charter, work plan, and risk management plan should be developed and applied throughout the project. More on that in Section 4.

If you're still deciding whether you should become a PM, make sure the job is for you. See Appendix C for a description of a great PM.

For those hiring PMs, please choose your person right – for their skills and personality – and be ready to back them along the way (See Factor 6. below regarding project sponsorship).

Example of Project Plan (Gantt Chart)

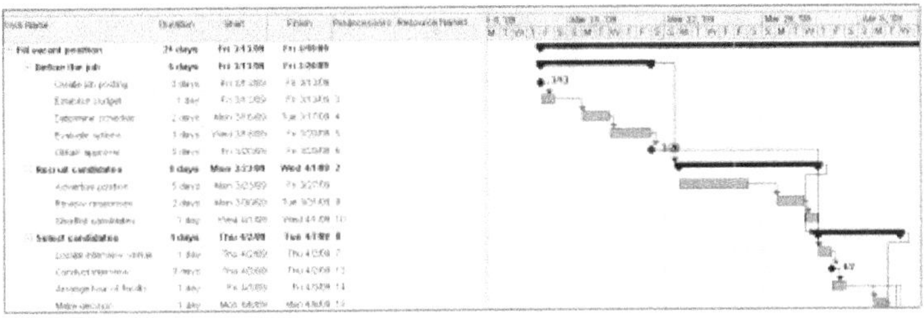

4- Poor communication & Inadequate Stakeholder Engagement

- ### *Why does this happen?*

Project managers may not be aware of, or yet trained to address, communication needs. They often focus on the tasks at hand and the resources needed to complete these tasks. This may result in an oversight of how the activities and outcomes of the project will impact various stakeholder groups beyond the immediate project team and what these stakeholders may need to know at each stage of the project.

A communication specialist is not always assigned to developing a communications plan, schedule, and/or key messages for the project either. Communications specialists work in tandem with the PM/team to craft content that's relevant to the subject matter and key audiences.

- ***What to do to prevent it Poor communication & Inadequate Stakeholder Engagement?***

Communication with team members, as well as with internal and external stakeholders, is critical to project success. You cannot deploy an information system or start a new program without telling people – i.e., anyone who will be impacted by the initiative – about the changes, and so forth.

Performing a thorough stakeholder analysis, developing a communication plan and key messages tailored to the different stakeholder groups, and communicating consistently throughout the project are key elements of a great communication and engagement strategy.

More on that in Section 5 and Appendix D.

5- Incomplete / Invalid Requirements

- ***Why does this happen?***

Requirements are considered part of the business analysis domain. Project managers, unless experienced and/or also trained in business analysis, are not familiar with the different types of requirements, let alone well-defined ones! So, they often just scope the project and use the scope as an input to their deliverables.

Good requirements cannot be developed if we don't know a) how things are done in the present ('as is') state, and b) what we want to change and what to in our future ('to be') state. This implies mapping business processes in the current environment. This also implies digging into data and workflows, especially if leading an IT project.

Inexperienced project managers may not know how else to proceed or to ask that a business analyst be assigned to develop proper requirements.

When a business analyst is not assigned to the project, team members (who are typically subject-matter experts and operational people, not project experts) are left to figure out what they need to do on their own. Without the proper analysis, requirements are not defined precisely and accurately enough to allow a clear understanding of the product(s)/service(s) expected. This results in outputs that do not quite meet what the client(s) had in mind. In instances when the team member responsible for a given deliverable decides to go back and forth to ensure the deliverable is right, a lot of re-work may ensue thus impact time, quality, and costs... And moods. There's nothing more frustrating for sponsors than being asked multiple times about the same thing – even when slight progress is made.

- ***What to do to prevent Incomplete / Invalid Requirements?***

First, depending on the size of the project, complexity of the deliverables, and your experience as a project manager, give some consideration to whether a trained business analyst should be added to your project.

Then, regardless of who is responsible for the gathering of requirements, schedule consultation sessions/workshops with team members and key stakeholders to elicit input into the mapping and documentation of processes that are within project scope. The next step, developing requirements, involves both work sessions with stakeholders and parallel reviews by subject matter experts.

During mapping sessions, make sure that every step pertaining to the processes that are in scope are defined and documented, and that pain points associated to these steps are identified and prioritized. Then, potential solutions to address them are elicited from those who do the work and validated/supported by managers/sponsors. With iterations and input by project specialists, the overall ideal solution will start emerging.

From there, requirements to elaborate the solution are written with the appropriate business analysis tools and templates.

Sample Requirements Traceability Matrix

Item ID	Requirement Description	Design	Development	Testing	Test ID	Deployment
RQ1	User Registration	Yes	In progress	Pending	T001	Pending
RQ2	Client Search (incl. filtering)	Yes	Completed	In progress	T002	Pending
RQ3	Referral Form	Yes	Completed	Completed	T003	In progress
RQ4	Notification to PCP	In progress	Pending	Pending	T004	Pending
RQ5	Appointment confirmation	Yes	In progress	In progress	T005	Pending

Note: *Requirements vary depending on the type of project. Watch out for software development and deployment projects where privacy and security requirements will also come into play, in addition to functional, non-functional, and technical requirements.*

6- Lack of Executive Support & Lack of resources

- *Why does this happen?*

Lack of Executive Support can mean a variety of things. On one hand, it can refer to the broader executive not supporting the project. While this may at times happen due to a lack of interest for a particular project, this may be avoided or minimized by ensuring that executives have signed off on it. The Factor 1. sub-section above illustrates how a pre-planning phase can bolster executive buy-in.

Unless they previously were in a project management role themselves or have already sponsored and learned about proper project execution along the way, executives and managers don't typically know what's expected of them as *Project Sponsors*. While Project Management and Change Management courses provide some information on what a good PS should do, they otherwise focus on teaching the practitioners – i.e., the project manager and change manager – on the methodology.

Further to that, sponsors may also not feel like the project-related risks and issues are not theirs to take on. As most intend to be in their positions for some time, or at least to leave with great references, it is much more fitting to have the 'good' role. Studies have shown that most people who got to a higher position made it by being nice – usually by managing up, but in some cases by also keeping their people/teams happy. Based on that alone, it makes sense that letting the PM handle the project, and be the 'bad guy' when needed, works well for such sponsors. *After all, that's why they hire contractors to execute projects, right?* Maybe it is in their mind, but while they're tending to everything else in their purview (i.e., the 'day-today' operation), the project may be suffering and not delivering what it was meant to.

For others who do not provide the necessary support, it may of no fault of their own. After all, executives usually run from meeting to meeting, often putting out one fire after another, and don't have the cycles to oversee every project in the way that best practices describe the project sponsor role.

- ***What to do to prevent a Lack of Executive Support and/or resources?***

If you are an executive/manager and was asked to be a PS, I highly recommend becoming familiar with what you are expected to be accountable for as a start, then evaluating whether you have the capacity to assume sponsorship for the project. You can be a 'Business' Sponsor without being the 'Executive' Sponsor. This involves providing resources and participating in key meetings but is less time consuming and doesn't usually require that you be available when something is awry with the project. If you must take executive sponsorship on, consider a) delegating some of your day-to-day to a trustworthy colleague or subordinate, and/or b) sharing the PS role with another person. While the later is not ideal, it can make sense to structure a project as such when two or more departments are impacted in the same manner by the project (i.e., they have equal stakes in it and should equally be entitled to making decisions as to its direction.

As for PMs, while it's not officially your responsibility to train managers to do their job as sponsor and you cannot force them to assume accountability for the project, I do advise coaching and educating them about their duties at onset of the project. These include (see Appendix E for details):

> Introducing / championing project across the organization

> Securing resources (Capital / Human)

> Removing barriers

> De-risking and addressing issues

> Attending key team meetings

> Meeting with PM on a regular basis to discuss progress

> Reviewing project documentation

> Signing off on key decisions and final deliverables

Overall, the sponsor's role is to show up for you and intervene when something is not going well. For those who do not believe in the ***servant leader*** philosophy, it is a little bit harder to do. But, don't get discouraged PMs, there is still a way to get your PS to be a good sponsor in spite of that. If they realize how much the time spent giving projects the right attention pays off, they might just make an extra effort!

– 1 –

You Should Get Pretty Familiar with the Project Knowledge Areas and Integrating All of Them

K nowing enough to manage each of the knowledge areas can save any sponsor or PM, especially if you don't have training or a certification in project management.

Eight (8) Knowledge Areas + Integration Management

Note: Activities for each knowledge area may occur before or after the project is formally approved. The below does not distinguish either way. Also, this section is not meant to provide an in-depth introduction to the Knowledge Areas. The PMBOK is the resource for this content.

For those who are certified, you'll notice that the PMP covers these in a different and more elaborate way. I opted to present them here in a simplified manner to hopefully make them easier to absorb. The purpose of this book is not to replace the PMBOK!

As mentioned earlier, I knew very little about project management when I started managing projects. My client took a chance on me because he saw that I had many transferable skills from my previous job as an airline call centre coach/trainer and that my MBA education had provided me with knowledge in key areas that would benefit my PM role. The later, my natural organizational and people skills, as well as my persistence to get results, are what got the job done for the two projects I was contracted to deliver.

Of course, the more educated and experienced you are as a PM, the more savvy you are expected to be in each of these areas. This is especially true if you wish to grow in your responsibilities and take on large projects.

The eight **Knowledge Areas** of the project world are:

1 – Scope Management

- **Project Scope** is:

> A detailed outline of all aspects of a project, including all related activities, resources, timelines, and deliverables, as well as the project's boundaries.

> Scope should answer: *What, When, Why, Where, Who, How.*

- **Scope management** is generally about:

> Properly defining what the project will and will NOT do/explore and deliver – so the In / Out of Scope) – at a decent level of specificity.

> Documenting the defined scope, and having project sponsor and team agree to it.

> Once project is approved, ensuring that all follow through on what was agreed-upon.

> Following the proper change procedures if an amendment to the scope is sought, and ensuring that required changes are made if approved.

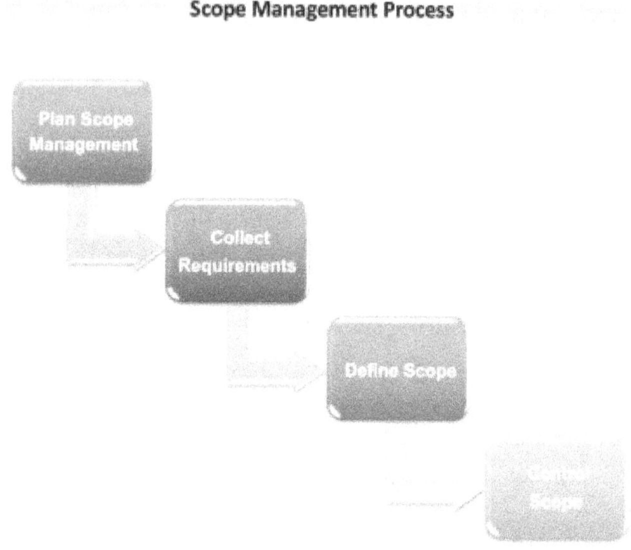

Following this process can prevent scope creep and a negative impact on all other dimensions of the project.

2 – Time Management

- **Project Time** refers to:

> The organization of time to complete project work, which is cyclical in nature.

> It is a component of the Magic Triangle and one of the central variables in project management.

- **Time management** is generally about:

> Clarifying how much time a project requires.

> Setting up a strategy to allocate the right amount of time to each task.

> Deciding on deadlines for project phases and delivery dates.

> Ensuring deadlines are known by team members and are adhered to.

> Updating the plan as task duration or effort changes are reported.

> Following the proper change procedures if an amendment to the set deadlines is sought, and obtaining approval.

Project time management directly impacts the quality, scope, and cost of a project, making it one of the most important project management knowledge areas. Managing time **helps securing project completion on time and on budget**.

3 – Cost Management

- **Project Cost** refers to:

> The total funds needed to monetarily cover and complete a business transaction or work project. Project costs involve direct costs.

> **Direct costs** are those necessary to complete said project.

- **Cost management** is about:

> Planning all of the resources needed for the project and estimating related costs.

> Securing and managing the project budget.

> Ensuring all are aware of the time they can allocate to the project and, as relevant, the amounts that can be spent to achieve their respective objectives.

> Reporting project spendings to keep teams on budget and overall costs reasonable.

> Updating the budget as needed.

> Following the proper change procedures if an amendment to the budget is required, and obtaining approval.

Project cost management directly impacts the quality, scope, and time that can be spent on a project, so it is also one of the most important project management knowledge areas. Managing the budget properly ensures that all activities and deliverables in scope can be completed as defined.

4 – Quality Management

- **Project Quality** refers to:

> The quality of outputs and the project's plans, procedures, and progress.

> Quality levels of the schedule, budget, performance, and client satisfaction must be met to be deemed successful.

- **Quality management** is generally about:

> Establishing and agreeing on the level of quality sought for each deliverable/outcome.

> Planning to perform quality assurance/control during the project.

> Ensuring that quality is indeed achieved and finding ways to accomplish agreed-upon quality levels more rapidly.

> Controlling the level of quality of each deliverable before it is submitted/signed off.

> Continually measuring the quality of all activities and taking corrective action until the team reaches the desired quality.

The Quality management process helps to control the cost of a project, establish standards to aim for, and determine steps to achieve standards.

Quality Management Process

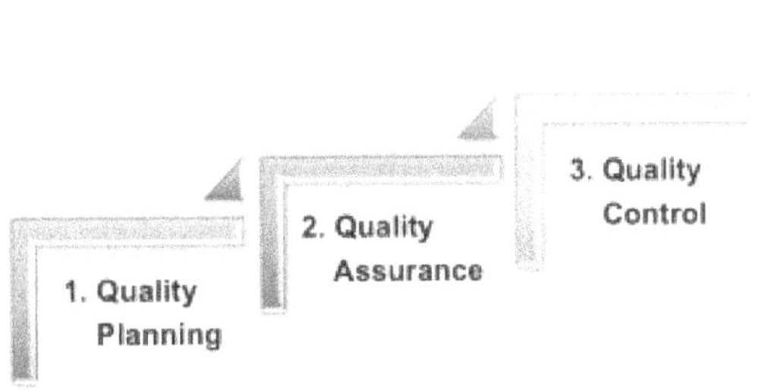

See Appendix F for more information on Quality Assurance & Control.

5 – Human Resource Management

- **Project Human Resource** is:

> The organization, management, and leading of a project team.

> The team includes everyone who has been assigned a role and responsibilities for completing the project. This may include internal (i.e., staff) and external resources such as suppliers and collaborators from partner organizations.

- **Human Resource management** is generally about:

> Planning the human resource requirements – including skills, experience, number of resources, hours allocation (e.g., FTE), and length of time required for each resource.

> Selecting resources for the project team and confirming their role/time on the project.

> Orienting everyone to the project in their respective roles, as well as building capacity amongst the team.

> Managing the team and releasing resources as/when required.

6 – Communication Management

- **Project Communication** is:

> The process of exchanging information and confirming there is shared understanding.

> Decisions about communication methods are made in the context of the target audience, the intended impact, and the risks/potential for unintended consequences of the approach.

- **Communication management** is generally about:

> Defining the audience (stakeholders) that have an interest in project activities and/or outcomes.

> Collecting and analyzing data relevant to stakeholders, including their potential interest and influence on the project.

> Determining the best method to provide information to stakeholders based on their preference, level of interest and influence.

> Developing key messages and other content to be transmitted or distributed in different formats.

> Scheduling and disseminating communications, including newsletter, presentations, and project FAQs.

> Following up with stakeholders requiring clarifications.

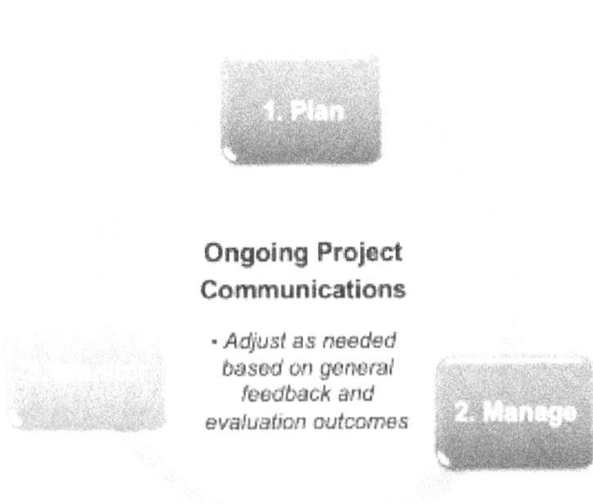

See Appendix D for more information on Communications Planning.

7 – Risk Management

- **Project Risk** is:

> The potential that a circumstance that could alter the outcome of a project, for better or for worse, arises.

> Project risks affect deliverables, timelines, and budgets. They can lead to project failure if not managed properly.

- **Risk management** is about:

> Registering all known project risks (Endnote 2) in a risk log.

> Assessing the probability each risk will materialize and become an issue, and the impact this would have.

> Preparing a response, and a contingency plan in the event the initial response does not alleviate the risk.

> On an ongoing basis, handling any risk that arises during the life of the project, so it remains on track and meet its goal.

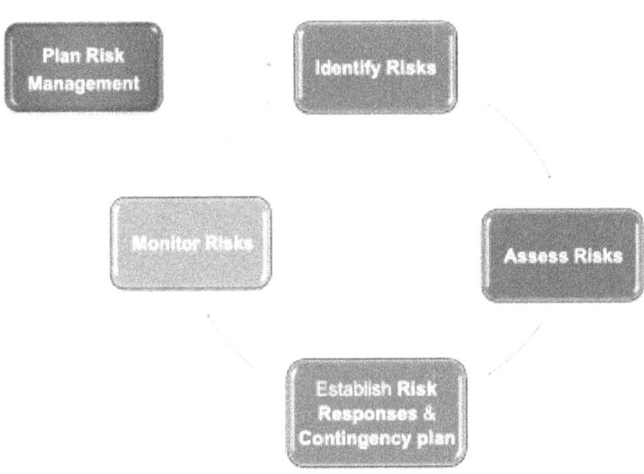

8 – Procurement Management

- **Project Procurement** is:

> The act of obtaining goods, supplies, and/or services – thus obtaining all of the materials and services required for the project.

> Project procurement management encompasses the processes used for making sure the procurement of resources for the project is successful.

- **Procurement management** is about:

> Defining procurement objectives.

> Developing the Procurement Strategy and Procurement Management Plan.

> Identifying potential suppliers, soliciting supplier proposals, selecting suppliers, and finalizing contracts.

> Monitoring supplier performance and verifying deliverables.

> Managing changes and resolving contract disputes, if applicable.

> Closing contracts and evaluating suppliers.

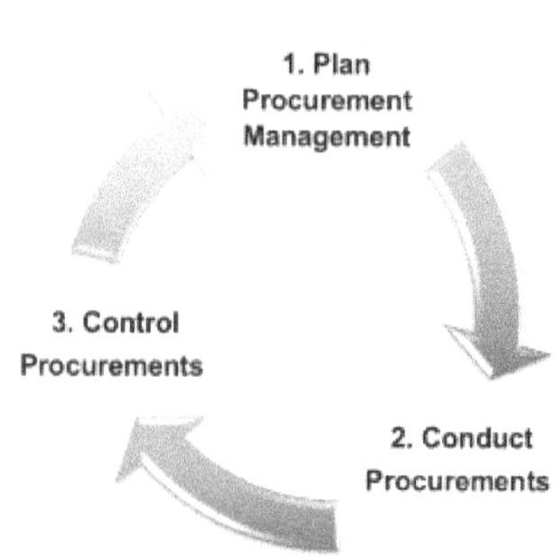

– 2 –

Knowing the Business Subject Matter Helps, but It Is/Should NOT be a Must

Depending on your background and the type of project management you are looking to do, you may wonder whether you should already have a certain expertise and experience in the subject matter. It's a fairly typical question.

For instance, because many project managers in IT were previously software engineers, there is a widespread perspective that project managers who lead software development or deployment projects should have an education or direct experience in software development. The same goes for construction projects – i.e., a background in construction or engineering (e.g., civil) is expected – or, for instance, clinical trials or drug approval projects – i.e., a background in the medical or pharmaceutical field is preferred. I could name many like this.

The thing is, we all must start somewhere and, even for IT projects, it doesn't have to be in engineering. After all, do we want all project managers to be cut from the same cloth? Engineer types may be highly performing on a technical level; however, a great strength of PMs comes from being a 'emotionally attuned to others'. Only a minority of engineers have both. (No offense to folks with engineering degrees – I really do admire your smarts and general perseverance!)

For government contracts, you will be faced with this requirement over and over: "X years of experience in Y type of projects", or yet "X number of projects delivering Z", where of course Y and Z match the subject matter of the project the government department is seeking to staff through a procurement process of some sort.

Eventually, as you progress towards more senior roles, you will be required to demonstrate that you've completed 'X number of projects' in the domain of expertise you are bidding/submitting for. This means that while the business or subject matter of the client is not as relevant per se, expertise in the type of project they are looking to implement is. For example, Public Health Agency of Canada (PHAC) is looking to upgrade their SAP financial and procurement system. Do you need to know about Public Health? No, it's a nice to have. But you will need to have done SAP implementations before. Now, if PHAC wishes to replace their legacy public health surveillance application, would you need to show a pretty good understanding of public health surveillance? You bet, you would! You would also need to understand, for instance, the types of events that would trigger alerts and notifications to interested parties. The way to do that is through previous experience in that field or through related education.

This may all seem to go against the message we hear during project management courses – a PM does not need to be a subject-matter expert (SME), as they have a team that comes with the complement of expertise and knowledge that is required to deliver. The PM is there to leverage the talents of each team member by organizing the work and ensuring progress is made. While that's theoretically true, it cannot always be replicated in reality. Government entities are accountable to the public ('taxpayers') for their consultants and suppliers' performance and must be able to justify why they chose a certain contractor versus another. Procurement processes (e.g., Request for proposals) requiring proven expertise through experience (and education) does that for them. That way, they can minimize the risks of project failure, shorten the learning curve of resources, and ultimately reduce costs by sourcing someone who not only has the project management experience and knowledge but also the subject-matter expertise – at least to some extent. You will also need to obtain a minimal security clearance (Endnote 3) to work on government contracts at the Federal level and to pass a background check for many others.

Unless you are set on government contracts, you should not be subjected to the same requirements by private companies or NGOs – at least not at the beginning of your career. Those organizations mainly want to know that you'll do a good job as a PM. (They will also pay you less.) You should be able to 'sell' your capabilities/competencies. Demonstrating that you have some or great knowledge in their area of business, especially in regulated industries, will go a long way. This will again become especially true as you grow more senior and command a higher fee/salary. The daily rate you can expect will vary based on your experience within and across industries.

Be mindful of being 'type casted' as a certain type of PM, with skills only in a specific area. As much as not specializing enough in an area of choice will limit you later in your career, overspecializing could also severely limit your opportunities – i.e., your ability to move laterally across different types of projects. Like everything in life, it's all about finding the right balance.

I recommend trying different types of projects and clients, being exposed to various subject matter, and learning everything you can along the way to keep a competitive edge. Take professional development courses beyond project management (many qualify for PDUs if you need to maintain your PMP). It will pay off.

What I encourage most, however, is to choose projects with a subject matter that interest you in and even are passionate about. This will not only make your work more enjoyable, it will also have a positive impact on the whole team and deliverables/outcomes. As a result, clients will be happier and you will build a reputation as an amazing PM.

– 3 –

Being Technical is Not Essential, But It Certainly Helps!

———

As mentioned above, I didn't have an engineering or some other background that's technical (Endnote 4) in nature and I was able to deliver several technology-enabled projects.

I did a natural incline towards technology though, and I think that's needed if you're going to take on mandates involving software deployment, or even with some components linked to a software application.

I've often referred to myself as a 'Business PM' to explain that I acted as a liaison, or bridge, between clients, project team members from the business and IT side, vendors, etc. I always put myself in a position to know just enough from a technical and technological perspective to:

a) communicate with IT in a way that we understood each other and could be sure that we were talking about the same thing (or not);

b) ensure that I could catch risks/issues that others may not foresee or want to share (so, be able to sniff out if they're trying to avoid getting in trouble if they had messed something up); and

c) verify that we were progressing in the right
direction – that is, delivering per client expectations
and the triple constraint.

To be able to do this properly for each project, I learned everything I needed to know about the software application being deployed and, where applicable, what it was to deliver at different stages of the project. How did I learn what I needed? I read a lot and asked a lot of questions.

I recommend you do the same, whether you are a business or a technical PM, as you should never assume that you know everything and that things have not changed since you last deployed a certain system. In the current technological and technical landscape, things change as quickly as rapid fire travels!

This includes reading background documentation, such as functional, non-functional, and technical requirements, solution and architecture design, privacy and security assessments, etc. Project managers are supposed to read and sign off on project documentation but, most often, they barely read it. I read everything that's sent my way end-to-end and made sure I understand what it means. When I wasn't sure, I Googled or asked my trusted SMEs. That's what they're there for.

I spent time with my team members and key stakeholders to go over our respective understanding of what the project was to deliver and tapped into their expertise to fill in my own knowledge bucket and gaps.

Bottom line is – you don't need to have a technical background, but make sure to add a few technical cords to your arc. It will come in very handy to earn the respect and trust of your technical team and to identify, understand and mitigate risks/issues faced by the project.

– 4 –

You Don't Need all of the Methodology and Templates to be Successful

Like many, I started managing projects with very little training and knowledge of project management. I had learned the basics of MS Project during my MBA (a lot of it also came instinctive to me), so I started with developing a plan (in Excel) then sought more information and templates that would help me carry the project from my colleagues at the health authority that gave me my first mandate.

This was a time when project management was still relatively new and the methodology was not broadly established. There weren't a lot of tools around, let alone project management software. The internet wasn't the infinite resource that it is today. Organizations were just starting to establish Project Management Offices (PMOs), and there was no consistency anywhere on how to manage projects.

From there, as I got more PM contracts, I researched project management methodology and tools and leveraged what I could use. I adapted existing templates and tools, and also developed my own along the way.

The PMP provided a framework and also referenced many templates and tools, but it didn't provide the templates. I was fortunate to have access to various sets through clients and, eventually, to many different styles online. I often got inspired and shamelessly borrowed from others' creative genius. But I also shared my own templates free of charge, in the course of paid engagements, over and over as a way to give back. Why re-invent the wheel when so many before us have developed and made available thousands and thousands of templates?

Unless you're planning to work on very large IT or real property projects (> $50,000,000), you will not need the full gambit of templates and tools. The Project lifecycle that is referenced in Appendix A is an intermediate version of the PM methodology and artifacts you might need to develop and deliver as part of an average size project. The larger the project, the greater the rigour that must be applied to ensure control is maintained on all the moving parts. This so they don't go off the rails!

With small projects, you should to a minimum have a work plan, a risk management process, and a reporting mechanism (e.g., project status updates via email and in a presentation on a PowerPoint deck). I'd also recommend a mini project charter and resource plan to inform the budget.

With a medium size project, additional artifacts such as a more comprehensive project charter, a communications plan, a requirements definition document, a formal status report (i.e., dashboard), a training plan, and a simplified version of a post-mortem with a lessons learned summary are advised.

With larger projects, you will need much more but you should also have the resources specializing in each area to help you develop and maintain that documentation.

Ultimately, being organized and having a mind for planning and structuring things logically will get you a long way.

– 5 –

Communicating & Managing Change Is Part of Your Job

—————

Communication is critical to project success. You cannot deploy an information system or start a new program without telling people – i.e., anyone who will be impacted – that it is happening. This includes properly and consistently communicating to internal and external stakeholders **what** the change is, **who** and **what** (processes, etc) it is going to impact and how, **when** stakeholders will be impacted (significant dates, not just the launch date), **why** it is happening, **what** will be gained from it, **how** changes will unfold, and so forth.

Many projects fail because no one was identified to develop the communication plan and ensure that related activities take place. It is often assumed to be the PM's responsibility since a 'Communications Specialist' is not always assigned to the project. In fact, I've observed that most clients' communications department are there to review communiqués, not to craft and/or disseminate project communications.

Performing a thorough stakeholder analysis, crafting a communication plan and key messages tailored to different stakeholder groups, and following a communication schedule are key elements of a great communication and engagement strategy. See Appendix D for details.

This is also true for **Change Management (CM)**. I'm talking about managing the 'people' vs 'technical' aspect of changes here. (Don't get me wrong, technical aspects must also be managed, but that falls into controlling the project and scope management.)

CM – like Business Analysis, Communications, Project Management, Software Engineering, etc – is its own discipline. It is separate from Project Management. However, to be effective, it requires the following:

1- effective communication;

2- alignment and/or integration with project activities;

3- participation/uptake by every member of the project team.

This means that you, as the face of the project, will need to assume at least some responsibilities for championing objectives so that others in the organization come along, and for supporting teams through the transition. The sooner they accept that things are changing, the faster and easier you will be able to progress with project activities.

I've applied CM principles within my PM practice for over 15 years. As a former internal consultant for a large health authority I, with fellow consultants, developed and facilitated a workshop to build project management capacity across the organization. We integrated CM principles to our PM methodology because we knew that it provided a solid approach to delivering projects. I continued doing so for my own projects throughout my career as I've observed the positive effects of it.

Change management is more than communication, but they work hand in hand. As PM who will be required to 'sale' the benefits of the project, engage with and seek endorsement from stakeholders, etc you will be performing a lot of the CM activities. Even if you always were to have a CM specialist assigned to your project (Endnote 5), you should at least know what their role is and why it's important.

Knowing this, you now have no excuse for your projects to fail based on poor communication or change management!

– 6 –

You are NOT a Project Coordinator

A lot of clients confuse project management with project coordination. This can make your job quite tedious and even frustrating.

While some of the responsibilities of the project manager are to coordinate the team, tasks, and some of the meetings to get the team moving, it should not be the PM's role to:

> manage minute administrative items such as stakeholder distribution lists, etc;

> schedule every meeting with broader stakeholder groups document/take notes at every meeting;

> maintain/proofread and distribute all of the project documentation

> upload and replace project documentation on a shared repository (such as SharePoint);

> manage and communicate basic information about project meetings, timesheets, etc.

The list goes on, but I hope you get my drift now.

The project manager is meant to lead, orchestrate the work, and oversee the overall project – the big picture – on behalf of the client. That's why they typically are a more costly resource. If they are bogged down with menial tasks, they cannot perform their role optimally – i.e., they are unable to keep an eye on all project activities/areas and anticipate what may go sideways, or plan to mitigate risks and address issues. They cannot be available to their sponsors and team members to provide support, nor proactively communicate more broadly about the project, or update/review project documentation.

Of course, junior project managers or other leaders that are responsible for small projects and smaller teams may need to perform a lot of the tasks that would otherwise fall to a Project Coordinator. It's actually the best way for them to understand what projects entail and subsequently guide their project coordinator when they have one.

Note that I am not saying that more senior PMs should NEVER perform these administrative tasks. I've completed many of these when resources were scarce, or everyone was busy and we all had to be hands on deck.

I've also done them when it just made sense to. Why getting my project coordinator to update one item in a document when I've got it open and can readily do it? Why getting my project coordinator to book every single meeting I have when it will take more time for me to go back and forth with them regarding meeting details and timing?

What I'm really saying is – you should not be expected to assume these tasks. Clients often have a misconception of what a PM does (they too often see us as 'coordinators'), so it is in your interest to clarify up front with your client the type of administrative/coordination support they expect you to handle in your new mandate.

If a project coordinator cannot be added to the team, there should at least be an administrator (executive assistant or else) to help you book meetings with your sponsor(s), broader stakeholder groups, and even to take meeting notes. I was very fortunate that my first project sponsors had this all lined up for me so I did not have to wait years to benefit from this. I've had project coordinators more often that not. I saw a huge difference when I did not, so it's worth fighting for it.

– 7 –

You Need to be Strategic, a Savvy Negotiator

Every project presents elements that need to be revisited or are yet to be decided. Because clients usually want it all (i.e., the best product, as fast as possible, and for the lowest price), PMs are constantly faced with managing the tension between each area of the triple constraint. The ability to think strategically, negotiate well, and even 'sale' the positive aspects of a certain approach versus another are invaluable assets to arrive at an agreement with the PS on what they can compromise on and to an understanding (or even acceptance) by team members and broader stakeholder groups that we must work within the set parameters (i.e., producing deliverables within the desired timelines, scope, and budget).

Thinking strategically is useful to devising approaches and tactics in every area of the project (e.g., communications & engagement, risk management), as well as for the overall execution of the project.

Many think that being 'strategic' is limited to the high spheres of business, but it also applies to projects. How? You are strategic when:

- At an *organizational* level, you choose the right projects for your organization – those that will put operations and people in a better position once executed.

> With resource limitations impacting how many initiatives an organization can undertake yearly, prioritizing is a must.

- From a ***mid-management*** standpoint, as an executive, you must state the outcomes (not the solution(s)), you want in a manner that your team(s) grasp. So, to be clear:

> The Exec Sponsor should dictate the WHAT (i.e., the results we want to achieve), and

> Project teams – with the PM's guidance – should figure out the HOW (i.e., how we're going to solve this, and how we're going to make it happen).

- At the ***project*** level, your role is to ensure your team understands the vision and connect project objectives to that big picture.

> If the vision is done right, project objectives will make sense and what the project delivers for the organization will too.

> The Sponsor can bridge the strategy to the tactical for and with the execution team, so that the Visionaries and other strategists don't need to go into the weeds.

Years ago, the Executive VP at the health authority where I worked stopped me in the hallway to ask what I was working on these days. I replied that I was doing a review of Relocation processes across our agencies and would soon have a report with recommendations on how 'moves' could be completed more cost-effectively and smoothly.

He said: "What? Your branch is supposed to take on strategic projects. That's not strategic!!" I replied that it was strategic in the sense that we'd define and apply a strategic solution to an operational problem that was grossly impacting departments across the health authority. He thought I was being silly. I was serious.

He was right that it wasn't strategic from a 'Strategic Plan' point of view. However, to me and per Harvard professor and strategy guru Michael Porter[1], being TRULY STRATEGIC is to align and integrate strategy with project management and reality – at all levels, and across the organization.

Both PS and PMs have the responsibility to convey that message through words and actions.

Use your selling skills to rally people who do not yet view the world that way.

---

1. https://www.hbs.edu/faculty/Pages/profile.aspx?facId=6532

– 8 –

Your Role is to Make Your Client(s) Look Good

As a PM, you must keep your client in mind. Your 'client' is not only the person who engaged you or the PS, but also those who work for them.

Whether you are an internal or external (contracted) resource, your role is to support these clients in meeting objectives for their departments/programs.

Depending on what is more important to you and whether you're looking to get repeat contracts from the same clients or not, you might tend to employ different strategies – e.g., being nice all the time and pushing less to ensure people like you and call you back to provide additional services, or being nice when people do their job and tough when they don't.

Either way, whether you are fussed about your contract being renewed or not, you will still have to play nice and to overcome your pride when dealing with people who are unpleasant or who treat you as if you are 'less than', or a 'passing and disposable' commodity.

You will also need to tame your pride to make your clients shine and achieve their objectives. This may also signify taking the hit or the high road when a client/your employer or a member of your team:

a) takes credit for your work;

b) blames you for an error that wasn't yours or a slippage impacting the triple constraint that was out of your control (e.g., you raised the risks or asked for changes and no one listened), or

c) messed something up and jeopardized project success.

The key is to be hard on the processes, not the people. This means keeping objectives in mind and focusing on how to get there. We are all human, and no one is perfect. Often people are dealing with difficult situations, and they cannot deliver well. Will getting angry and snappy at them help the situation? Maybe in the very short-term, but that's it.

If you state expectations at onset, share required information, encourage team members to contribute to components of the project structure where feasible, anticipate and strategically mitigate risks, proactively manage deliverables and timelines, coach junior team members, and remove barriers with the support of the PS, then you have done what you could. It is likely not your fault if something goes sideways. It may not be your team member's fault either. Either way, addressing it in an accusatory way is not the way to go. Give your people the benefit of the doubt and try to understand what may have led to something being late or delivered with a lower level of quality than what was expected. Something out of the team member's control may have occurred that resulted in such a situation. Most people want to do well and also want to ingratiate themselves to their bosses, and that means make them look good. So be supportive and help them get through the tough patch.

If the same issue occurs over and over with different people, you will see that the problems stem from a defective process. If the issue is repeated by the same and only this individual, then you will need to consider gently removing them from the project.

Remember: If your team doesn't look good, you client doesn't look good. And no one looks good if the team is constantly experiencing conflict, or there is a consistent and tangible tension. That inevitably takes away from delivering quality work within the agreed timeline and even budget.

– 9 –

Signing Off Does NOT Just Mean 'Adding Your Signature'

M any documents are produced and circulated during a project.

As a PS, you are responsible for reviewing every deliverable that comes your way for signature or sign off, and accountable for the content that the team has developed and is proposing to operationalize. There can be a significant impact to the final results and outcomes, so to your department and organization's Key Performance Indicators (KPIs), if you signed off on something that was flawed, incomplete or misaligned with other important components of your programs.

If you do not understand certain parts of the deliverable, let your PM know that you have some questions. If your PM is unable to explain or to make sense of it in a way that satisfies you, ask them to organize a walkthrough with the SMEs who injected content into the deliverable. It is possible that the way they drafted it was not clear or quite in alignment with specifications and, as a result, the document requires editing. Thus, such discussions can lead to a greater understanding and comfort on your part that you can sign off, or to improved final deliverables. Either way, this will strengthen the due diligence around the project.

As a PM, it is also your responsibility to review every deliverable – especially those that will be shared with your client(s) and/or the public – and to ensure that designated SMEs and business owners also review and sign off on each deliverable before submitting to the PS, who is ultimately accountable for accepting all project deliverables.

Inexperienced PMs often omit to build time into the plan for reviews and quality assurance by key members of the team and sponsors, including themselves. They may underestimate the time that will be required by each team member to review each deliverable. This should never be overlooked – because even when the estimated effort is low, the duration might actually end up being three or ten-folds the effort. This will take the plan off track if the time wasn't properly built in. Given the number of meetings on people's calendars these days, there is less free time to 'do the work' and even much less to review documents sent from multiple sources.

This is why I highly recommend the following:

   a) incorporating review cycles into your project plan & schedule (including planning for several review cycles as there are always items requiring clarifications or several cycles of edits)

   b) identifying all individuals you expect to sign off

   c) appointing someone who is excellent at editing and proofreading to ensure no errors remain in the document from both content and grammar standpoints

d) creating a deliverable traceability matrix and/or review schedule with all identified for yourself or your project coordinator to track

f) communicating with those who are expected to provide content and complete reviews by:

1. including team members and key stakeholders in the development and validation of the workplan so they can inform the steps and turnaround time required for each deliverable;
2. once completed, sharing the plan with all who are expected to contribute to deliverables, so they know when they will be expected to contribute/submit;
3. using an electronic workflow/sign off application such as DocuSign or Adobe signature or SharePoint Collect Signatures.

As PM, it is your role to ensure that all of these activities are duly managed and completed.

Of course, you can delegate their coordination to your PC if you have one.

– 10 –

You May Need to be the 'Bad Guy' to Stick to the Triple Constraint

While this section addresses the PM, this also very much applies to the current and future PS reading this book. I hope that you will consider how much the PM might bear on their shoulders when a PS doesn't fulfill their role and provide support, and also the optic this gives off to others around.

As stated earlier, a PM needs to remain focused on project objectives and the elements of the triple constraint. This implies getting comfortable with being the bad guy on occasion (for some projects, on more than one occasion).

Of course, different projects and organizations have different priorities and levels of elasticity with how much a project can derive from set objectives/constraints.. But, really, what's the point of setting certain goals and then not sticking to them?

Make sure your Sponsors are clear about this. Setting a line in the sand that will keep on moving is a bad look for them as decision-makers and for you too as project lead. It indicates to the project teams that the PS don't know what they want, are pushovers, can't be trusted, or yet that they (the team) don't have to listen to you because your Sponsor will keep overriding you anyway.

I once had a team member ask me, a few weeks before our final proposal was due: "So, why does it have to be finished that week? Is this an absolute?". The deadline had been set several months before, at the beginning of the project, by our Executive Sponsor (his boss twice removed), and we had discussed several times during team meetings why this timeline was set. I knew the reason for him asking wasn't that he didn't understand the rationale, nor that he'd forgotten, nor that he wasn't listening in meetings and didn't hear the rationale. No, it was that he had left the work he had to do on the proposal to the last minute, despite my multiple encouragements and reminders, and he didn't want to have to put in the extra effort to deliver on time. He had complained about his workload to his supervisor and to our PS (they were a tight-knit team) and the later, a people pleaser who was often wishy-washy, had told him to see with me – as I managed the plan and the deadlines. She took no accountability for the deadline. I had to be the tough one and stick to the timeline that our PS had set, or it would defeat the purpose of the project altogether. That is, we would miss the window we had to submit the proposal and see another budget cycle (i.e., a full year) pass before government could move with any of the recommendations. The guy didn't like it one bit, but he delivered. Was he sad to see me go at the end? No. Are we still in touch? No. But that is the reality I chose.

As hard as it can be at times, you'll need to be the bad guy or play devil's advocate and apply some rigour to ensure objectives can be attained within the constraints, even when you don't agree or feel like it. To be able to deliver to the level clients expect for high priority or visibility projects, you need to demand and push people to follow a methodology and deliver on aspects that many don't enjoy.

On one hand, there are always team members who have never been on a project and are unfamiliar, even uncomfortable, with that structure. You may have to 'educate' more than you anticipated, want to, or have time to do. So, interestingly, the change management you have to do is often not just with supporting people with the changes that will come as a result of the work done on the project, but also about helping team members adapting to the project management methodology and the structure (e.g., meeting, status reporting) you are establishing for your particular project.

On the other hand, since most people don't like to be reminded of deadlines or told their work is not up to standards – even by their own boss, imagine how it goes over when done by a complete stranger! That alone explains the resistance PMs get from staff who are on their project team.

You will need to use all the tools in your arsenal to finely balance a supportive, yet firm, approach.

I'll always remember what my first manager answered when I expressed surprise that a colleague had decided to leave her position as a corporate coach/trainer to take on an operations manager role. I thought our unionized staff would eat her alive, but our manager knew her differently. She said: "She'll be fine. She has an iron rod under that velvet glove." PMs need that. You must approach people gently and, while you shouldn't have to use an iron rod, you need to make them feel like there will be real consequences if they don't follow through on their responsibilities and your requests. (I mean, if there are no consequences to not finishing on time, on budget and delivering something of quality, then what's the point of having a PM leading that project to begin, or even to executing that project at all?!!)

Being a PM, like with most things, is a balancing act – in this case, between showing support, but also challenging and expecting from your team that they deliver.

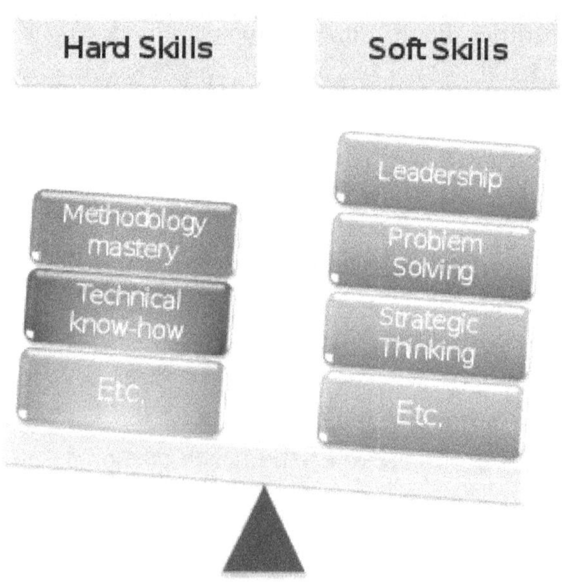

APPENDICES

(Tip: Click on the images to increase the size & font.)

Appendix A – Project Lifecycle

The project lifecycle below illustrates the key activities, deliverables (outputs), and approvals required at phase.

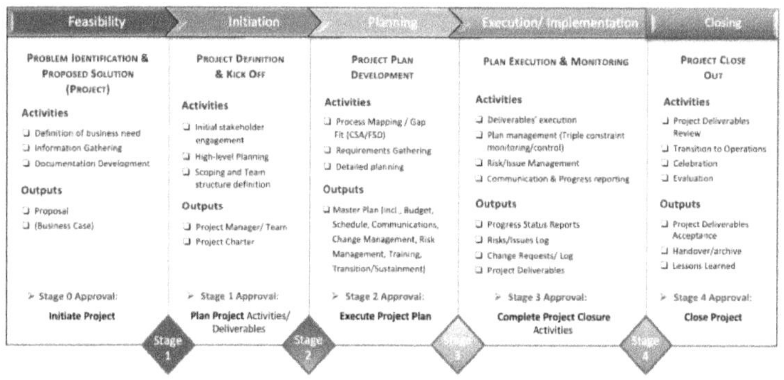

[Back to top](#)

Appendix B – Triple Constraint Variations

The Triple Constraint model below shows how balancing each corner of the triangle should result in delivering overall quality for the project.

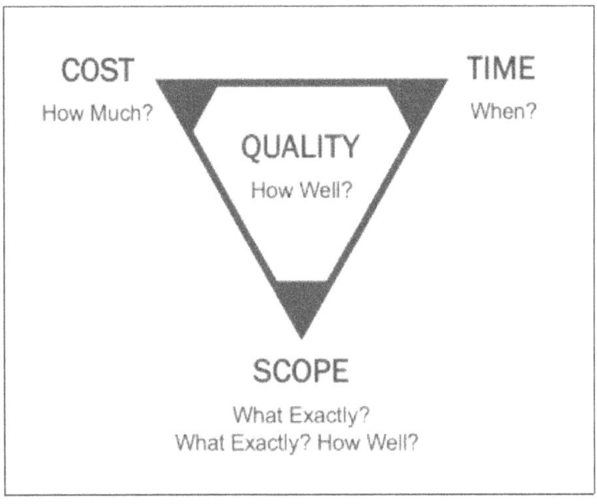

This Triple Constraint model illustrates the tension between cost, time, and performance (i.e., scope & quality).

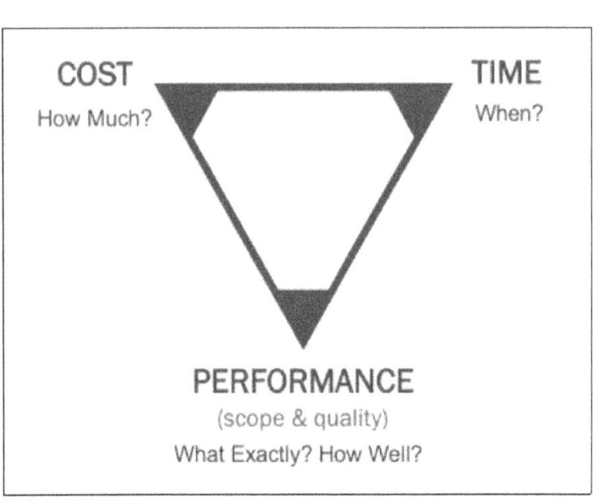

You cannot have it all (i.e., a product with all the bells and whistles, in the shortest timeframe and at the lowest price).

[Back to top](#)

Appendix C – What Makes a Great Project Manager?

While there is no formal code of conduct for project managers (like the one accountants, doctors, lawyers, management consultants, and other professional streams with designations have), there are still expectations for PMs to adhere to certain ethical and professional standards. Behaving with integrity and following best practices are some examples of such standards. But, beyond that, what makes a PM great at their job?

Great PMs are both **process and goal oriented**. They ensure that the right processes are in place and standardized to get a) get to end goal, b) ensure that project outcomes are sustained following go live. They wear multiple hats as:

- **Subject-matter experts (SME)** – in the PM methodology and learning/knowing enough about the subject matter concerning the project at hand.
- **Doers** – in getting things done in their role as a Project Manager (executing the vision, putting the structure in place for the project, managing the team, etc), without doing the work for team members.
- **Motivators** – encouraging team members to do the work and others to support the vision and embrace the change.
- **Delegators** – Assign tasks that should not fall to

them to others.

Above and beyond expected PM responsibilities, great Project Managers successfully joggle three main areas of leadership:

- **Coaching** – Educate and Motivate
- **Communicating** – Engage, Inform, Market
- **Delivering** – Walk the talk, and get it done!

While wearing these different hats, great PMs:

- ***Challenge and Believe***

- Finding the right balance, they incite their team to push themselves within their capacity and capabilities, and to demonstrate how much they believe in them to deliver on project objectives.

- ***Inspire a Shared Vision***

- They speak of the vision as it is a common goal – that belongs to the whole team – not just 'their' goal (as in the organization's, the Sponsor's, or the PM's). This motivates teams.

- ***Enables Others to Act***

- They find solutions and ways to remove barriers and support teams to complete their tasks in any way they can.

- ***Model the Way***

- They are a good example of how to behave in a project environment and apply best practices, principles, and values.

- ***Encourage the Heart***

- They persuade by appealing to the team's emotions. Anything done without passion or even interest is never well done. And what's the point of doing something if not done right?

<u>Back to top</u>

Appendix D – Communications & Engagement

- **Stakeholder Engagement & Communications Planning**

The following diagram lays out the key steps of Communications and Stakeholder Engagement planning. While a communications plan can be developed without a full stakeholder analysis, its quality will be lessened if the team does not take the time to properly identify and prioritise all project stakeholders, and gauge how they might impact the project if they are not engaged in a certain way or extent.

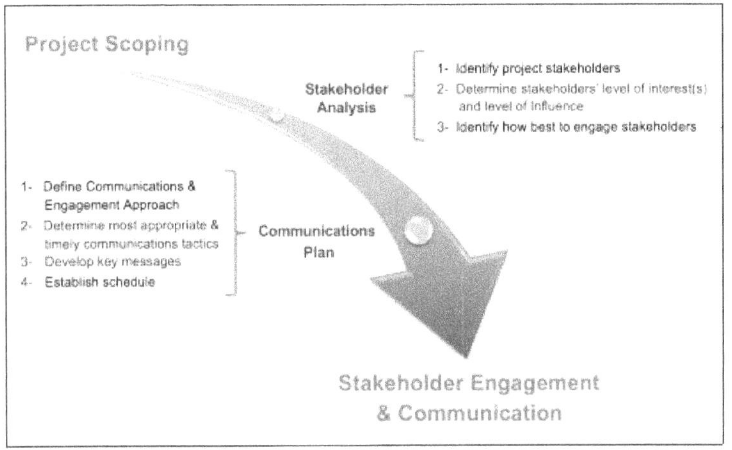

- **Stakeholders Analysis**

Many tools exist to help you complete a Stakeholder Analysis, to track stakeholders and how they fit into the project. You can easily create (or find online) a spreadsheet to do so.

One tool you might not as easily find or know how to use that I find very useful to assess stakeholders is the Stakeholder matrix:

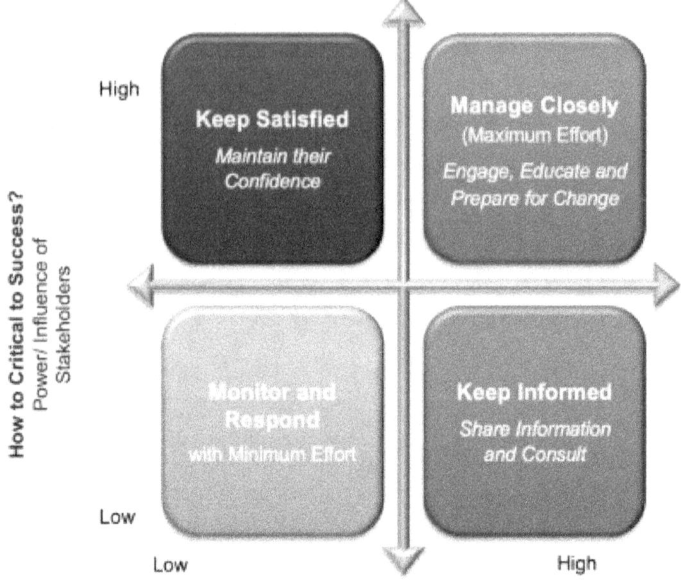
High
Keep Satisfied
Maintain their
Confidence
Manage Closely
(Maximum Effort)
Engage, Educate and
Prepare for Change
How to Critical to Success?
Power/ Influence of
Stakeholders
Monitor and
Respond
with Minimum Effort
Keep Informed
Share Information
and Consult
Low
Low
High
How Much Effort is Needed?
Interest of Stakeholders

It tells you clearly, based on their level of interest and (potential) influence on the project, how much effort you should invest in keeping them informed and involved.

One way to arrive at this is, working with team members, business owners and other key individuals, is to plot the different stakeholders identified in the quadrant using a 'blank' version of the matrix – e.g.,

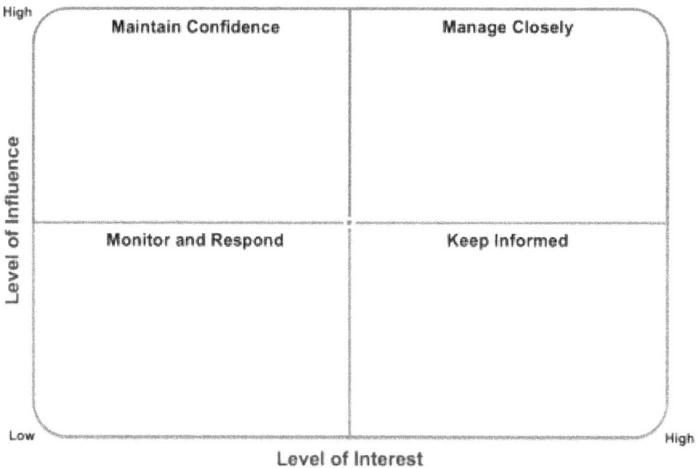

Back to top

Appendix E – What Makes a Great Executive Sponsor?

A great Executive Sponsor is someone who has a keen interest in the project's progress and outcome, as well as the capacity to give it the attention the project requires – from both a time and skills perspectives.

Great Executive Sponsors are successful at:

- **Introducing and championing the project across the organization**

- They make a point of introducing the project at every level of the organization at existing/standing meetings and in specially- scheduled meetings (e.g., kick offs) and speaking of project elements that will benefit the group. They should also not shy away of discussing how implementation will impact people's work.

- **Securing resources** (Capital / Human)

- They advocate for the resources needed to support the project and seek assignment of these resources for the duration of the project.

- **Removing barriers**

- They make sure that PMs and their staff have a clear path to move forward.

- **De-risking and addressing issues**

- They look for ways to eliminate risks as they are identified by PM/teams and do not shy away from facing and managing issues.

- **Attending key project meetings**

They make the time in their calendar to attend meetings where they are required – whether to get updates, to provide input, or take anything back for a later decision.

- **Meeting with PM on a regular basis to discuss progress**

- They make time for their PM. Even when there at no decisions, the PM might be bringing information forward that Sponsors – with their *organizational view* – may either find relevant for other purposes or can flag as a potential risk or issue for this or another project.

- **Reviewing project documentation** (in a *timely* fashion)

- They go through every part of documents that are relevant to their role. Good documents have

Executive Summaries that concisely describe key points and a table of content with hyperlinks to each section to allow the reader to navigate to sections of interest.

- **Signing off on key decisions and final deliverables**

- They pay attention to requests made by the PM/business leads for their decision and sign off on project phases and deliverables upon review. Without sign off, the team cannot move to the next stage so omitting these steps can lead to significant delays.

Overall, outstanding sponsors show up and support the PMs and team members. They also adopt the servant leader's mantra and, as such, exemplify how much they trust and value the PM and project team, and put the spotlight on them.

<u>Back to top</u>

Appendix F – Quality Assurance & Control

The following diagram illustrates the elements or deliverables involved with each step of the **Quality Assurance & Quality Control** process:

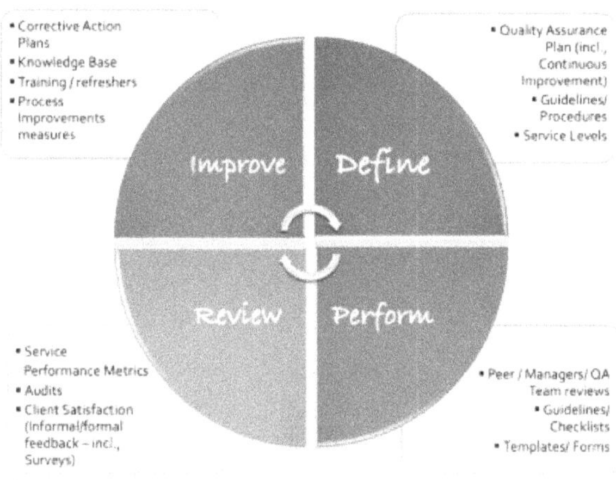

<u>Back to top</u>

Definitions

- **Agile***:* A method of project management used especially for software development, that is characterized by the division of tasks into short phases of work and frequent reassessment and adaptation of plans.
- **Artifacts:** A document designed to keep the project work aligned to project requirements and business goals. The PMBOK® Guide – 7th edition defines a project artifact as: "a template, document, output, or project deliverable."
- **Business Analysis***:* Business Analysis is the practice of enabling change in an organizational context, by defining needs and recommending solutions that deliver value to stakeholders.
- **Business Client**: A client, often also referred to as customer, can be a person or an organization that orders and buys products or services that a business offers. In project management, the customer is the one defining the requirements of the project and often setting the parameters such as budget and deadlines.
- **Business Owner (AKA Business Sponsor):** An advocate for the business representatives on the project team. There can be multiple business owners (BOs) depending on the complexity of the project and the nature of the deliverables. Working closely

with the Executive Sponsor, BOs define the vision and roadmap, and communicate to stakeholders. They have the knowledge and authority to make strategic decisions and clear the path of political and financial obstacles.

- **Change Management (CM)***:*

1. The management of change and development within a business or similar organization.
2. The controlled identification and implementation of required changes within a computer system.

- **Duration***:* The *total amount of time it takes to finish a project*, which you measure in business days, hours, weeks, or months. The duration can be seen at the timeline for project delivery, whether it is five days or five years. Project duration usually depends on your resource availability.
- **Effort***:* Time needed to complete a task, activity, or project. Effort refers to the number of labor units required to complete a task, activity, or project, and are often called 'man-hours'. Effort is usually expressed as time units such as days, hours, and minutes.
- **Executive Sponsor (AKA Project Sponsor)***:* A C-level senior member of a project management team with for duties to ensure a project's goals are aligned with the overall organizational strategy, gather support, communicate goals and overcome resistance from senior executives, in addition to provide ongoing direction to the project team during a project's

lifecycle. Project managers typically report to the Sponsor (or a Project Lead).

- **FTE:** Full-time employee (1 FTE = 1 F/T employee; 0.5 FTE = ½ time employee). Often used to establish project resource requirements.
- **Frequently Asked Question (FAQ):** A question in a list of questions and answers intended to help people understand a particular subject. If you have any problems, consult the FAQs on our website.
- **Issue:** A point or matter in question or in dispute, or a point or matter that is not settled and is under discussion, or over which there are opposing views or disagreements.
- **Key Performance Indicators (KPIs):** A quantifiable measure of performance over time for a specific objective. KPIs evaluate the success of an organization and/or its activities.
- **Lessons Learned:** A session that focuses on identifying project success and failures, and includes recommendations to improve future performance on projects. Also see Post-mortem.
- **PDU:** Per the PMI, Professional Development Units (PDUs) are one-hour blocks of time that an individual holding a PMP certification spends learning, teaching others, or volunteering. By accumulating and tracking these over a three-year certification period, one can maintain their certification status with PMI.
- **PMBOK:** Project Management Body of Knowledge (PMBOK) is *a set of standard terminology and*

guidelines (a body of knowledge) for project management.

- **Post-Mortem**: A project review session held through a (series of) meeting(s) once the project is completed. This is a chance for the project team to identify what went well throughout the project and what could have gone differently. Also see Lessons Learned.

- **Pre-Planning Phase:** This phase includes the development of the concept of the project, including the basic decision of selecting the concept that will be used for the execution of work.

- **PRINCE2® (Projects IN Controlled Environments)**: A process-based methodology for effective project management, used and recognised all over the world. It can be scaled and tailored to each project.

- **Project Constraints**: Any restriction that defines a project's limitations. Key project management constraints are time, cost, and scope.

- **Project Coordinator**: Within the broader scope of project management, a project coordinator organizes and manages the various parts of a project to ensure its success. This includes assigning and monitoring daily tasks (e.g., scheduling meetings, taking notes) and communication, as well as creating reports and updates for the project manager and other members of management.

- **Project Lifecycle:** Framework comprising a set of distinct high-level stages (i.e., initiation, planning, execution monitoring & control, closing) required to transform an idea of concept into reality in an orderly and efficient manner.

- **Project Management**: The discipline of initiating, planning, executing, controlling, and closing the work of a team to achieve specific goals and meet specific success criteria. It is the application of knowledge, skills, tools, and techniques to project activities to meet the project requirements.

- **Project Management Institute (PMI)**: A not-for-profit professional membership association for project managers and program managers. PMI was started in 1969 and now has a membership of more than 2.9 million professionals worldwide.

- **Project Management Professional (PMP)®**: The PMP certification is a globally recognized project management certification that tests a candidate's ability to manage the people, processes, and business priorities of a professional project.

- **Project Requirements**: The specific goals, objectives, functionalities, and features that a project must meet or possess to be considered successful. These requirements are typically outlined at the beginning of a project and serve as a guide throughout the project's lifecycle. Typical types of requirements are:

- **Non-functional requirement (NFR)**: Requirement that specifies criteria that can be used to judge the operation of a system, rather than specific behaviours. The plan for implementing *non-functional* requirements is detailed in the system architecture, because they are usually architecturally significant requirements.

- **Functional requirements** : Define specific behavior or functions. The plan for implementing *functional* requirements is detailed in the system design.

- **Privacy requirements:** Requirements of an organization, information program, or system that are derived from applicable laws, Executive Orders, directives, policies, standards, instructions, regulations, procedures, or organizational mission and business case needs with respect to privacy.

- **Security requirement**: A statement of needed security functionality that ensures one of many different security properties of software is being satisfied. Security requirements are derived from industry standards, applicable laws, and a history of past vulnerabilities.

- **Technical requirements:** The technical issues that must be considered to successfully complete a project. These can include aspects such as performance, reliability, and availability. In software projects, technical requirements typically refer to how the software is built, for example: The programming language.

• **Project Sponsor:** A project sponsor (PS) is

responsible for the overall success of a project by providing financing and supportive resources. Depending on the complexity of the project, there may be an "Executive Sponsor" and one or more "Business Sponsor(s)". While they are all considered project sponsors, the executive sponsor is ultimately accountable for project outcomes. Project managers typically report to a Project Sponsor.

- **Risk**: An uncertain event or condition that, if it occurs, has a positive or negative effect on the project objectives.

- *Known Risks* can be identified, analyzed and planned for.

- *Unknown Risks* can be addressed by applying a general contingency based on past experiences.

- **Scope:** The defined features and functions of a product, or the scope of work needed to finish a project. Scope involves getting information required to start a project, including the features the product needs to meet its stakeholders' requirements. PMI applies narrow version of this definition while others use broader version. Then, it stands that per the Project Charter and Change Requests:

- *In-Scope* is what the project commits to do or include in its activities and/or deliverables.

- ***Out of Scope*** is what the project agrees <u>not</u> to include in its activities and/or deliverables.

- **Scope Creep:** Adding additional features or functions of a new product, requirements, or work that is not authorized (i.e., beyond the agreed-upon scope).

- **Servant Leader**: Leader who focuses on the needs of others before they consider their own. It's a longer-term approach to leadership, rather than a technique that can be adopted in specific situations. Therefore, it can be used with other leadership styles such as Transformational Leadership.

- **SMART (objectives):** SMART stands for Specific, Measurable, Achievable, Relevant, and Time-Bound. Defining these parameters as they pertain to your goal helps ensure that your objectives are attainable within a certain time frame.

- **Subject Matter Experts (SMEs)***:* A person who has accumulated great knowledge in a particular field or topic and this level of knowledge is demonstrated by the person's degree, licensure, and/or through years of professional experience with the subject.

- **Stakeholder:** A person with an interest or concern in something, especially a business.

- **Internal stakeholders:** People whose interest in a company comes through a direct relationship, such as employment, ownership, or investment.

- **External stakeholders:** Those who do not directly work with a company but are affected somehow by the actions and outcomes of the business.

- **Stakeholder analysis**: A process of identifying all stakeholders before the project begins, grouping them according to their levels of participation, interest, and influence in the project, and determining how best to involve and communicate with each of the stakeholder groups throughout.
- **System Architecture**: The conceptual model that defines the structure, behaviour, and more views of a system. An architecture description is a formal description and representation of a system, organized in a way that supports reasoning about the structures and behaviors of the system.
- **Systems Design**: The interfaces and data for an electronic control system to satisfy specified requirements. Systems design could be seen as the application of systems theory to product development. There is some overlap with the disciplines of systems analysis, systems architecture, and systems engineering.
- **Systems Engineering:** An interdisciplinary field of engineering and engineering management that focuses on how to design, integrate, and manage complex systems over their life cycles. At its core, systems engineering utilizes systems thinking principles to organize this body of knowledge. The individual outcome of such efforts, an engineered

system, can be defined as a combination of components that work in synergy to collectively perform a useful function.

- **Triple Constraint:** Infers that the success of the project is impacted by its costs, time, and scope. The Project manager can keep control of the triple constraint by balancing these three constraints through trade-offs.

<u>Back to top</u>

References & Resources

O Forbes Advisor[1]: Forbes, amongst the multitude of business services it provides, issues regular (monthly) articles on different project management topics.

O International Institute of Business Analysis[2]: Official website of the IIBA that offers certifications and resources in business analysis.

O KPMG Survey[3]: KPMG is a trustworthy and world-renowned accounting and *consulting firm that* offers professional services, including project management, for businesses across industries.

O PM Illustrated[4]: A website describing the full project lifecycle with images and videos to support concepts!

O Project Management 101[5]: In-person and virtual training offered by PMC Training[6], which covers

---

1. https://www.forbes.com/advisor/search/?q=project%20management

2. https://www.iiba.org/

3. https://assets.kpmg.com/content/dam/kpmg/cy/pdf/2023/

   kpmg_pmi_project_management_survey_2022.pdf

4. https://www.pmillustrated.com/

5. https://pmctraining.com/training/

   project-management-101/#_6666cd76f96956469e7be39d750cc7d9_

the fundamentals and best practices of project management.

O Project Management Institute[7]: Official website of the PMI that offers certifications and resources in project management.

O Projectmanager.com[8]: Project management software website that also provides useful resources on project management.

O *Integrated Project Management* workshop – B.C. Provincial Health Services Authority: Three-day training using best practices of change management, process improvement and project management developed and delivered by the Strategic Planning and Transformation Support department.

O Simplilearn[9]: An online courses and certifications provider, it also offers resources in project management and other domains.

O Wikipedia[10]: A free online encyclopedia, created and edited by volunteers around the world and

---

6. https://pmctraining.com/?utm_source=redirect301&utm_source=website-link&utm_medium=redirect&utm_medium=organic&utm_campaign=gbp#_6666cd76f96956469e7be39d750cc7d9_

7. https://www.pmi.org/learning

8. https://www.projectmanager.com/

9. https://www.simplilearn.com/resources#project-management

10. https://en.wikipedia.org/wiki/Main_Page

hosted by the Wikimedia Foundation. It contains a wealth of knowledge!

Endnotes

—

(1) Several sources, citing studies by IBM, KPMG, and the Project Management Institute, indicate that around 60% of projects fail in one area or more. This number reaches 70% for IT projects.

(2) Some risks that were not identified may also rise as issues. This is called "unknown unknowns", as in 'the risk was unknown to the project team, and the consequence on the project was also unknown'.

(3) There are several levels, including Secret and Top secret, but most contracts require Enhanced Reliability. This is not too difficult to obtain if your record is clean, but it may take time to get it the first time you apply.

(4) I here include security and pure technological aspects in technical.

(5) More often than not, organizations misjudge the role of the CM consultant and have this role report to the PM. While a CM can report to a Project Executive, the CM consultant should have a direct line to the Executive Sponsor to ensure that key players are engaged through Sponsor's effort.

Back to top

About the Author

Julie Barré is a public sector consultant and former project manager, the proud mother of a pre-teen girl and a Yorkipoo, and a blooming philanthropist.

She is hoping to pass on as much of her knowledge and build capacity amongst the existing and new generation of Project Sponsors and PMs before retiring!

Read more at https://www.jube-consulting.com.